How Quantum Physics Explains Everyday Phenomena

Mike Bhangu

Published by BB Productions
British Columbia, Canada
thinkingmanmike@gmail.com

How Quantum Physics Explains Everyday Phenomena

This book is designed to make quantum physics accessible and relatable, using everyday experiences as a lens to explore complex concepts. Each chapter will blend science, storytelling, and practical insights to engage readers.

Table of Contents

Introduction: The Quantum World in Your Everyday Life

Hook: The Quantum in the Quotidian

Imagine this: you wake up to the sound of your alarm, brewed coffee waiting in your smart mug, and your phone buzzing with notifications. You glance at your GPS to check the fastest route to work, all while marveling at how your wireless earbuds seamlessly connect to your playlist. What you might not realize is that every one of these everyday conveniences relies on the strange and fascinating principles of quantum physics. From the transistors in your phone to the lasers in your Wi-Fi router, the quantum world is not just the domain of scientists in lab coats—it's the invisible force shaping your daily life.

Thesis: Quantum Physics is Everywhere

This book is not about equations, lab experiments, or abstract theories. It's about you. It's about how the seemingly bizarre rules of quantum physics—superposition, entanglement, uncertainty, and more—are not just confined to the realm of subatomic particles. They are playing out in your relationships, your decisions, your struggles, and your triumphs. By the end of this journey, you'll see that quantum physics isn't just a science; it's a lens through which to understand the world—and yourself—in a whole new way.

Why Quantum Physics Matters to You

You might be thinking, "Quantum physics? That's way over my head." But here's the thing: you don't need a PhD to appreciate its beauty or its relevance. Quantum physics is already a part of your life, whether you realize it or not. It's in the way your morning coffee stays hot (thanks to the quantum properties of heat transfer), the way your GPS knows exactly where you are (relying on atomic clocks and relativity), and even the way your brain processes information (with quantum-like decision-making). This book will show you how these concepts apply to the mundane, the ordinary, and the deeply human.

A Beginner-Friendly Approach

This book is designed for the curious, not the experts. You won't find complex equations or jargon here. Instead, we'll use relatable stories, analogies, and examples to make quantum physics as intuitive as your morning routine. Each chapter will explore a key quantum concept and connect it to an everyday experience, from decision-making to relationships to personal growth. By the end, you'll not only understand quantum physics—you'll see how it explains the world around you.

What You'll Learn

Here's a sneak peek at the quantum concepts we'll explore and how they relate to your life:

-Superposition: How you exist in multiple states (e.g., "Should I hit snooze or get up?") until you make a decision.

-Entanglement: How your relationships defy distance and logic, much like particles connected across space.

-Uncertainty: Why life is unpredictable and how to embrace the unknown.

-Wave-Particle Duality: The many roles you play in life and how you adapt to different contexts.

-Quantum Tunneling: How you can break through barriers and achieve the seemingly impossible.

-The Observer Effect: How your attention and focus shape your reality.

-Quantum Coherence: The power of alignment in your goals, relationships, and teams.

-Entropy: The natural tendency toward chaos—and how to create order in your life.

-Quantum Leaps: Moments of sudden change and transformation.

-The Multiverse: The parallel lives and possibilities that exist in your imagination.

The Promise of This Book

By the end of this book, you'll see the world differently. You'll notice the quantum in the mundane—the invisible forces shaping your decisions, your relationships, and your experiences. You'll gain a deeper appreciation for the complexity and beauty of everyday life, and you'll walk away with practical insights to apply these principles to your own journey. Whether you're a science enthusiast or a complete novice, this book will leave you with a sense of wonder and a new way of thinking about the world.

How to Use This Book

This book is designed to be read in any order. Each chapter stands on its own, so you can jump to the topics that intrigue you most. However, if you're new to quantum physics, starting from the beginning will help you build a foundation for the concepts that follow. Along the way, you'll find:

-Real-world examples: Stories and analogies to make abstract concepts tangible.

-Practical takeaways: Insights you can apply to your own life.

-Questions for reflection: Prompts to help you connect the ideas to your experiences.

Let's Begin

Quantum physics might seem like a world of paradoxes and mysteries, but it's also a world of endless possibilities. It's a reminder that the universe is far stranger—and far more wonderful—than we often realize. So, let's take a step into the quantum world together. By the time you finish this book, you'll see that the mundane is anything but ordinary. It's quantum.

This expanded introduction sets the stage for the book, making quantum physics approachable, relatable, and exciting for readers. It invites them on a journey of discovery, promising both intellectual stimulation and practical insights.

Chapter 1: Schrödinger's Morning Routine— Superposition and Decision-Making

The Paradox of the Snooze Button

It's 7:00 a.m., and your alarm blares. You're faced with a choice: hit snooze and steal nine more minutes of sleep, or get up and start your day. For a moment, you exist in a state of limbo—neither fully asleep nor fully awake. In that brief window, you're in a "superposition" of two states, much like Schrödinger's famous cat, which is simultaneously alive and dead until someone opens the box. Your morning routine, it turns out, is a perfect example of quantum superposition in action.

What Is Superposition?

At its core, superposition is the idea that a quantum system can exist in multiple states at once until it is observed or measured. In the quantum world, particles like electrons don't have definite properties until we look at them. They exist in a cloud of possibilities, a blend of all potential outcomes. This might sound bizarre, but it's a fundamental principle of quantum physics—and it has surprising parallels in our everyday lives.

Superposition in Everyday Life

Think about the decisions you make every day. From the moment you wake up, you're faced with choices: what to wear, what to eat, which route to take to work. Each decision involves weighing multiple possibilities, and until you make a choice, you're in a kind of superposition. You're not just one version of yourself—you're all the potential versions of yourself, each tied to a different outcome.

-Example 1: The Breakfast Dilemma

You stand in front of the fridge, torn between a healthy smoothie and a comforting stack of pancakes. Until you decide, both options exist in a superposition. The act of choosing "collapses" the possibilities into a single reality.

-Example 2: The Career Crossroads

You're offered a new job opportunity. Do you stay in your current role, where you're comfortable but stagnant, or take the leap into the unknown? Until you decide, you're in a superposition of two futures.

The Quantum Mechanics of Decision-Making

In quantum physics, the act of observation collapses the wave function, forcing a particle to "choose" a state. Similarly, in life, the act of making a decision collapses your superposition, bringing one possibility into reality while the others fade away. This process can feel daunting—after all, every choice means saying no to countless alternatives. But it's also empowering. Just as particles navigate their quantum states, you have the power to shape your reality through your decisions.

The Role of Uncertainty

One of the key insights of quantum physics is that uncertainty is built into the fabric of the universe. You can't know everything at once—a principle known as the Heisenberg Uncertainty Principle. This applies to decision-making, too. You can't predict the outcome of every choice, and that's okay. Embracing uncertainty allows you to explore possibilities without being paralyzed by the fear of making the "wrong" decision.

Practical Takeaways: How to Navigate Superposition

1.Embrace the Process

Instead of rushing to make decisions, allow yourself to sit with the uncertainty. Just as particles exist in a cloud of possibilities, you can explore your options without immediately committing to one.

2.Trust Your Intuition

In quantum physics, observation collapses the wave function. In life, your intuition can act as your "observer," helping you make decisions that align with your values and goals.

3.Celebrate the Multiverse of You

Every decision creates a new version of you. Instead of regretting the paths not taken, celebrate the richness of your experiences and the person you've become.

A Deeper Dive: Schrödinger's Cat and You

Schrödinger's famous thought experiment involves a cat in a box that is simultaneously alive and dead until someone opens the box and observes it. While this seems absurd when applied to a cat, it's a powerful metaphor for the choices we face. Until we make a decision, we exist in a state of potential—a blend of all possible outcomes. The act of choosing brings clarity and direction, collapsing the possibilities into a single reality.

Closing Thought: The Beauty of Superposition

Superposition reminds us that life is full of possibilities. It's a call to embrace the uncertainty, to see the beauty in the "not yet," and to recognize that every decision is an opportunity to shape your reality. So the next time you're faced with a choice—whether it's hitting snooze or getting up, choosing pancakes or a smoothie—remember: you're not just making a decision. You're navigating the quantum landscape of your life.

Questions for Reflection

1. What's a recent decision where you felt "stuck" in superposition? How did you finally collapse the possibilities into a choice?

2. How do you handle uncertainty in your life? Can you think of a time when embracing the unknown led to a positive outcome?

3. If you could peek into the "multiverse" of your life, what alternate versions of yourself would you be most curious to meet?

This chapter uses the concept of superposition to reframe decision-making as a natural and empowering process. By connecting quantum physics to relatable experiences, it helps readers see the beauty and potential in their everyday choices.

Chapter 2: Entangled Relationships—How Connections Defy Distance

The Invisible Thread

Imagine this: you're miles apart from a loved one, yet you feel an inexplicable connection. You think of them, and moments later, they call or text. Or perhaps you've experienced a moment of shared understanding with a friend, where words weren't necessary—you just "knew" what the other was thinking. These moments of deep connection might feel like magic, but they have a surprising parallel in the quantum world:entanglement.

What Is Quantum Entanglement?

Quantum entanglement is one of the most mysterious and fascinating phenomena in physics. When two particles become entangled, they are deeply connected, no matter how far apart they are. Change the state of one particle, and the other instantly changes, too—even if they're on opposite sides of the universe. This "spooky action at a distance," as Einstein called it, defies our everyday understanding of space and time. But what if entanglement isn't just a property of particles? What if it's also a metaphor for the connections we share with others?

Entanglement in Human Relationships

Just as entangled particles are linked across vast distances, human relationships often defy logic and physical separation. Think about the bonds you share with family, friends, or romantic partners. Even when you're apart, you can feel their presence, their influence, and their emotions. This chapter explores how the principles of quantum entanglement can help us understand the invisible threads that tie us together.

The Science of Emotional Entanglement

1.Mirror Neurons and Empathy

Neuroscience has shown that our brains are wired for connection. Mirror neurons, for example, fire both when we perform an action and when we see someone else perform it. This neural mirroring is the basis of empathy—the ability to feel what others feel. In a sense, our brains are "entangled" with those of the people around us.

2.The Power of Intuition

Have you ever had a gut feeling about someone's mood or intentions, even without obvious cues? This intuitive connection might be a form of emotional entanglement, where your subconscious picks up on subtle signals and aligns with the other person's state.

3.Long-Distance Relationships

Even when separated by miles, people in close relationships often report feeling deeply connected. This could be seen as a kind of emotional entanglement, where the bond transcends physical distance.

Real-Life Examples of Entanglement

-The Parent-Child Bond: A mother wakes up in the middle of the night, sensing that her child needs her—only to find out later that the child had a nightmare at that exact moment.

-Twin Telepathy: Twins often report knowing what the other is thinking or feeling, even when they're far apart.

-Shared Grief or Joy: When a community comes together to mourn a loss or celebrate a victory, the collective emotion creates a powerful sense of connection.

The Quantum Mechanics of Connection

In quantum physics, entanglement occurs when particles interact and become linked, sharing a single quantum state. Similarly, in human relationships, connection is forged through shared experiences, emotions, and intentions. Just as entangled particles remain connected regardless of distance, the bonds we form with others can endure time and space.

Practical Takeaways: How to Strengthen Your Entanglements

1.Cultivate Presence

Just as observation plays a role in quantum entanglement, being fully present with others strengthens your emotional connections. Put away distractions and truly listen.

2.Practice Empathy

Empathy is the human equivalent of entanglement. Try to see the world from another's perspective and feel what they feel.

3.Nurture Your Relationships

Like entangled particles, relationships require energy and attention to maintain their connection. Make time for the people who matter most.

4.Trust the Invisible Threads

Even when you're apart from loved ones, trust that the bond you share is real and enduring. Distance doesn't diminish true connection.

A Deeper Dive: Entanglement and the Nature of Reality

Quantum entanglement challenges our understanding of reality, suggesting that the universe is far more interconnected than we realize. This interconnectedness isn't just a scientific concept—it's a profound truth about human relationships. We are all part of a vast, intricate web of connections, and the bonds we form with others are as real and powerful as the forces that hold atoms together.

Closing Thought: The Beauty of Entanglement

Entanglement reminds us that we are never truly alone. The connections we share with others transcend physical boundaries, creating a tapestry of relationships that shape who we are. Whether it's a loved one across the globe or a friend in the next room, the bonds we form are a testament to the interconnectedness of all things. So the next time you feel that invisible thread pulling you toward someone, remember: it's not just in your mind. It's quantum.

Questions for Reflection

1. Have you ever experienced a moment of deep connection that felt like entanglement? What was it like?

2. How do you maintain emotional connections with loved ones who are far away?

3. Can you think of a relationship in your life that feels "entangled"? What makes it so strong?

This chapter uses the concept of quantum entanglement to explore the profound and often mysterious nature of human relationships. By drawing parallels between the quantum world and our emotional lives, it helps readers appreciate the invisible threads that bind us together and offers practical insights for nurturing those connections.

Chapter 3: The Uncertainty Principle—Why Life Is Unpredictable

The Unpredictable Commute

You wake up early, plan your route to work, and leave with plenty of time to spare. But then—traffic. An accident on the highway, a delayed train, or an unexpected detour throws your carefully crafted schedule into chaos. No matter how much you plan, life has a way of surprising you. This unpredictability isn't just a quirk of modern living; it's a fundamental feature of the universe, as described by one of the most famous principles in quantum physics:Heisenberg's Uncertainty Principle.

What Is the Uncertainty Principle?

Formulated by Werner Heisenberg in 1927, the Uncertainty Principle states that there are inherent limits to what we can know about a particle's properties. Specifically, the more precisely we measure a particle's position, the less precisely we can know its momentum, and vice versa. This isn't a limitation of our tools or technology—it's a fundamental property of nature. The universe, at its core, is uncertain.

Uncertainty in Everyday Life

Just as particles exist in a state of inherent uncertainty, so do our lives. No matter how much we plan, prepare, or predict, there will always be elements beyond our control. This chapter explores how the Uncertainty Principle can help us understand and embrace the unpredictability of life.

The Science of Uncertainty

1.Quantum Fluctuations

At the quantum level, particles don't follow strict, deterministic paths. Instead, they exist in a cloud of probabilities, constantly fluctuating between states. This inherent randomness is a reminder that uncertainty is built into the fabric of reality.

2.Chaos Theory

In the macroscopic world, small changes can lead to large, unpredictable outcomes—a phenomenon known as the "butterfly effect." A butterfly flapping its wings in Brazil could, in theory, set off a chain of events leading to a tornado in Texas. This sensitivity to initial conditions makes long-term prediction nearly impossible.

3.Human Behavior

Even in our own lives, uncertainty reigns. From the stock market to the weather to the choices of the people around us, there are countless variables that we can't fully predict or control.

Real-Life Examples of Uncertainty

-Career Paths: You might have a clear plan for your career, but unexpected opportunities—or setbacks—can change everything.
-Relationships: No matter how well you know someone, their actions and decisions can still surprise you.

-Health: Even with advances in medicine, health outcomes are often uncertain. A diagnosis, an accident, or a sudden recovery can upend your expectations.

The Quantum Mechanics of Uncertainty

In quantum physics, the Uncertainty Principle isn't a flaw—it's a feature. It reflects the probabilistic nature of the universe, where particles don't have definite properties until they're observed. Similarly, in life, uncertainty isn't something to be feared or avoided. It's a natural part of existence, a reminder that the future is always open to possibility.

Practical Takeaways: How to Embrace Uncertainty

1.Focus on the Present

Just as particles exist in a cloud of probabilities, your life is a series of unfolding moments. Instead of fixating on the future, focus on what you can control right now.

2.Cultivate Flexibility

Uncertainty requires adaptability. Learn to pivot when things don't go as planned, and see unexpected changes as opportunities rather than obstacles.

3.Let Go of Perfection

The Uncertainty Principle reminds us that perfect knowledge is impossible. Accept that you can't predict or control everything, and give yourself permission to make mistakes.

4.Find Comfort in Probability

While you can't predict the future, you can prepare for it. Think in terms of probabilities rather than certainties, and make decisions based on the best information available.

A Deeper Dive: The Role of the Observer

In quantum physics, the act of observation plays a crucial role in determining a particle's state. Similarly, in life, your perspective and mindset can shape how you experience uncertainty. By adopting a curious, open-minded approach, you can transform uncertainty from a source of anxiety into a source of wonder and possibility.

Closing Thought: The Beauty of Uncertainty

Uncertainty is often seen as a negative—a sign of chaos or lack of control. But the Uncertainty Principle invites us to see it differently. It's a reminder that the universe is dynamic, creative, and full of potential. Just as particles exist in a state of possibility, so do we. The future isn't fixed; it's a canvas waiting to be painted. So the next time life throws you a curveball, remember: uncertainty isn't a flaw. It's a feature.

Questions for Reflection

1. What's a recent experience where uncertainty led to an unexpected outcome? How did you handle it?

2. How do you typically respond to uncertainty—with fear, excitement, or something else? How can you shift your mindset to embrace it?

3. Can you think of a time when uncertainty opened the door to a new opportunity or perspective?

This chapter uses the Uncertainty Principle to reframe unpredictability as a natural and even beautiful aspect of life. By connecting quantum physics to everyday experiences, it helps readers see uncertainty not as a threat, but as a source of potential and growth.

Chapter 4: Wave-Particle Duality—The Many Sides of You

The Chameleon in the Room

Think about the last time you switched roles in a single day. Maybe you started as a focused professional in a morning meeting, then shifted into a nurturing parent at school pickup, and later became a relaxed friend at dinner. Each version of you felt authentic, yet distinct. This ability to adapt and embody different roles isn't just a social skill—it's a reflection of a fundamental principle in quantum physics:wave-particle duality.

What Is Wave-Particle Duality?

Wave-particle duality is one of the most intriguing concepts in quantum physics. It states that particles, such as electrons and photons, can behave both as particles (discrete, localized entities) and as waves (spread-out, oscillating patterns). This dual nature depends on how they're observed or measured. For example, light acts like a wave when passing through a prism but behaves like a particle when hitting a solar panel. This duality challenges our classical understanding of reality and offers a powerful metaphor for the multifaceted nature of human identity.

Wave-Particle Duality in Everyday Life

Just as particles can exhibit both wave-like and particle-like behavior, humans can embody different roles, traits, and identities depending on the context. You might be a leader at work, a caregiver at home, and a dreamer in your private moments. These aren't contradictions—they're different facets of your complex, dynamic self. This chapter explores how wave-particle duality can help us understand and embrace the many sides of who we are.

The Science of Duality

1.Quantum Superposition

Before a particle is observed, it exists in a superposition of all possible states. Similarly, before you step into a specific role, you carry the potential for many identities.

2.Context-Dependent Behavior

Just as particles behave differently depending on how they're measured, humans adapt their behavior based on their environment and relationships. This adaptability is a survival mechanism, allowing us to thrive in diverse settings.

3.The Role of Observation

In quantum physics, the act of observation collapses the wave function, determining whether a particle behaves as a wave or a particle. In life, the way others perceive you—and the way you perceive yourself—can shape which aspects of your identity come to the forefront.

Real-Life Examples of Duality

-The Working Parent: At work, you're focused, decisive, and goal-oriented. At home, you're nurturing, patient, and emotionally attuned. Both roles are authentic, yet they require different sides of you.

-The Introverted Extrovert: You might thrive in social settings, drawing energy from others, but also need solitude to recharge and reflect.

-The Creative Analyst: You can switch between logical, analytical thinking and intuitive, creative problem-solving, depending on the task at hand.

The Quantum Mechanics of Identity

Wave-particle duality reminds us that reality isn't fixed—it's fluid and context-dependent. Similarly, human identity isn't a single, static thing. It's a dynamic interplay of roles, traits, and potentials. Just as particles exist in a state of possibility until observed, your identity is shaped by the situations you encounter and the choices you make.

Practical Takeaways: How to Embrace Your Duality

1.Celebrate Your Complexity

Instead of seeing your different roles as conflicting, recognize them as complementary. Each side of you brings unique strengths and perspectives.

2.Adapt with Intention

Like particles adjusting their behavior based on context, you can consciously choose which aspects of yourself to emphasize in different situations. This isn't about being inauthentic—it's about being adaptable.

3.Integrate Your Identities

While duality is a natural part of life, integration is key to wholeness. Find ways to connect your different roles and traits, creating a cohesive sense of self.

4.Embrace the Unknown

Just as particles exist in a state of potential, you are always evolving. Allow yourself to explore new roles and identities without fear of contradiction.

A Deeper Dive: The Double-Slit Experiment

The double-slit experiment is a classic demonstration of wave-particle duality. When particles are fired at a barrier with two slits, they create an interference pattern characteristic of waves. But when observed, they behave like particles, passing through one slit or the other. This experiment highlights the profound role of observation in shaping reality—and serves as a metaphor for how our self-perception and the perceptions of others influence our identity.

Closing Thought: The Beauty of Duality

Wave-particle duality teaches us that reality is richer and more complex than it appears. It's not a matter of either/or—it's both/and. You are not just one thing; you are many things, each one real and valid. So the next time you feel pulled between different roles or identities, remember: you're not fragmented. You're multifaceted, like a diamond reflecting light in countless directions. Embrace your duality, and let it shine.

Questions for Reflection

1. What are the different roles or identities you embody in your life? How do they complement or challenge each other?

2. Can you think of a time when you felt torn between two sides of yourself? How did you navigate that tension?

3. How can you integrate your different roles and traits to create a more cohesive sense of self?

This chapter uses the concept of wave-particle duality to explore the multifaceted nature of human identity. By drawing parallels between the quantum world and our personal lives, it helps readers appreciate the complexity and beauty of their many sides, offering practical insights for embracing and integrating their diverse roles and traits.

Chapter 5: Quantum Tunneling—Breaking Through Barriers

The Impossible Leap

Imagine standing at the base of a towering mountain, tasked with reaching the other side. The path is blocked by an impassable wall, and climbing over it seems impossible. But what if you could simply "pass through" the wall? This isn't the stuff of science fiction—it's a real phenomenon in the quantum world calledquantum tunneling. And while you can't literally walk through walls, the principles of quantum tunneling can inspire you to break through the barriers in your own life.

What Is Quantum Tunneling?

Quantum tunneling is a phenomenon where particles pass through barriers that, according to classical physics, should be impossible to penetrate. For example, an electron can "tunnel" through an energy barrier it doesn't have enough energy to overcome. This isn't magic—it's a consequence of the probabilistic nature of quantum mechanics. Particles don't have definite positions; they exist as waves of probability, and there's always a small chance they'll appear on the other side of a barrier.

Quantum Tunneling in Everyday Life

Just as particles can tunnel through barriers, humans can overcome obstacles that seem insurmountable. Whether it's breaking free from a limiting belief, achieving a seemingly impossible goal, or finding a creative solution to a problem, quantum tunneling offers a powerful metaphor for resilience and innovation. This chapter explores how the principles of quantum tunneling can help you break through the barriers in your life.

The Science of Tunneling

1. Wave Functions and Probabilities

In quantum mechanics, particles are described by wave functions that represent the probability of finding them in a particular location. Even when a barrier seems impenetrable, there's always a non-zero probability that the particle will appear on the other side.

2. Energy Barriers

In classical physics, a particle needs enough energy to overcome a barrier. But in the quantum world, particles can "borrow" energy from the uncertainty principle, allowing them to tunnel through.

3. Applications in Technology

Quantum tunneling isn't just a theoretical curiosity—it's the basis for technologies like flash memory, tunnel diodes, and even nuclear fusion in stars.

Real-Life Examples of Tunneling

-Overcoming Adversity: Think of someone who has risen above incredible challenges, like a refugee building a new life in a foreign country or an athlete recovering from a career-ending injury.

-Creative Problem-Solving: Sometimes, the best solutions come from thinking outside the box—tunneling through conventional wisdom to find a new approach.

-Personal Growth: Breaking free from limiting beliefs or habits often feels impossible, but with persistence and creativity, you can tunnel through to a new way of being.

The Quantum Mechanics of Resilience

Quantum tunneling reminds us that barriers aren't always as solid as they seem. Just as particles can pass through seemingly impenetrable walls, you can find ways to overcome the obstacles in your life. This doesn't mean ignoring reality or denying the difficulty of your situation—it means recognizing that there's always a possibility, however small, of breaking through.

Practical Takeaways: How to Tunnel Through Barriers

1.Embrace Uncertainty

Just as particles rely on the probabilistic nature of the universe to tunnel through barriers, you can embrace uncertainty as a source of possibility. Don't let fear of failure stop you from trying.

2.Think Outside the Box

When faced with a barrier, look for unconventional solutions. Sometimes, the path forward isn't over or around the obstacle—it's through it.

3.Leverage Small Probabilities

Even if the odds are against you, remember that small probabilities can lead to big outcomes. Keep trying, and you might just tunnel through.

4.Build Momentum

In quantum tunneling, particles don't stop at the barrier—they keep moving forward. Similarly, maintaining momentum can help you push through challenges.

A Deeper Dive: The Role of Persistence

Quantum tunneling isn't a guarantee—it's a possibility. The same is true for overcoming barriers in life. Success often requires persistence, creativity, and a willingness to take risks. Just as particles don't give up when they encounter a barrier, you can keep pushing forward, even when the odds seem stacked against you.

Closing Thought: The Beauty of Tunneling

Quantum tunneling is a reminder that the impossible is often just a matter of perspective. What seems like an insurmountable barrier might be a gateway to new possibilities. So the next time you're faced with a challenge, remember: you don't have to climb over the wall. You can tunnel through it.

Questions for Reflection

1. What's a barrier you've faced in your life? How did you overcome it—or how are you working to overcome it?

2. Can you think of a time when you achieved something that seemed impossible? What strategies did you use?

3. How can you apply the principles of quantum tunneling to a current challenge in your life?

This chapter uses the concept of quantum tunneling to inspire readers to break through the barriers in their lives. By drawing parallels between the quantum world and personal resilience, it offers a fresh perspective on overcoming obstacles and achieving the seemingly impossible.

Shapes Reality

The Power of a Glance

Imagine you're in a crowded room, and you suddenly feel someone's eyes on you. You turn, and sure enough, someone is staring at you. That simple act of observation changed your behavior—you noticed, you reacted, and the dynamic of the room shifted. This isn't just a social phenomenon; it's a reflection of a fundamental principle in quantum physics:the observer effect. In the quantum world, the act of observation doesn't just measure reality—it shapes it.

What Is the Observer Effect?

The observer effect is the idea that the act of observing a system inevitably changes that system. In quantum physics, this is most famously illustrated by the double-slit experiment, where particles behave differently when they're being observed. When no one is looking, particles act like waves, creating an interference pattern. But when observed, they act like particles, passing through one slit or the other. This suggests that reality isn't fixed—it's influenced by how we interact with it.

The Observer Effect in Everyday Life

Just as observation changes the behavior of particles, the way we focus our attention can shape our experiences and outcomes. Whether it's the placebo effect in medicine, the Hawthorne effect in workplace studies, or the self-fulfilling prophecy in psychology, the observer effect is a powerful force in human life. This chapter explores how the principles of the observer effect can help us understand and harness the power of attention.

The Science of Observation

1.Quantum Measurement

In quantum physics, measurement collapses the wave function, determining a particle's state. Similarly, in life, the act of observing or focusing on something can bring it into sharper reality.

2.The Placebo Effect

When patients believe they're receiving treatment, their symptoms often improve—even if the treatment is a placebo. This demonstrates how expectation and attention can influence physical outcomes.

3.The Hawthorne Effect

In workplace studies, employees change their behavior when they know they're being observed. This highlights the impact of attention on performance and motivation.

Real-Life Examples of the Observer Effect

-Self-Fulfilling Prophecies: If you believe you'll fail at a task, you're more likely to perform poorly. Conversely, if you believe you'll succeed, you're more likely to excel.

-Parenting: Children often behave differently when they know their parents are watching. This can be used to encourage positive behavior and discourage negative behavior.

-Mindfulness and Meditation: By focusing your attention on the present moment, you can change your mental and emotional state, reducing stress and increasing well-being.

The Quantum Mechanics of Attention

The observer effect reminds us that reality isn't a fixed, objective thing—it's shaped by how we interact with it. Just as particles change their behavior when observed, our experiences and outcomes can change based on where we focus our attention. This has profound implications for how we live our lives, from the goals we set to the relationships we nurture.

Practical Takeaways: How to Harness the Observer Effect

1.Focus on What Matters

Just as observation collapses the wave function, focusing your attention on what's important can bring it into reality. Set clear goals and prioritize your time and energy accordingly.

2.Cultivate Positive Expectations

The placebo effect shows that belief can influence outcomes. Approach challenges with a positive mindset, and you're more likely to succeed.

3.Be Mindful of Your Attention

The observer effect isn't just about external observation—it's also about self-awareness. Pay attention to your thoughts and emotions, and you can shape your inner reality.

4.Use Observation to Influence Behavior

Whether you're a parent, a manager, or a friend, your attention can influence the behavior of others. Use this power wisely, encouraging positive actions and attitudes.

A Deeper Dive: The Double-Slit Experiment

The double-slit experiment is a classic demonstration of the observer effect. When particles are fired at a barrier with two slits, they create an interference pattern characteristic of waves. But when observed, they behave like particles, passing through one slit or the other. This experiment highlights the profound role of observation in shaping reality—and serves as a metaphor for how our attention can shape our experiences.

Closing Thought: The Beauty of Observation

The observer effect is a reminder that we are active participants in our reality, not passive observers. The way we focus our attention, the expectations we hold, and the beliefs we nurture all shape the world around us. So the next time you feel stuck or overwhelmed, remember: you have the power to change your reality simply by changing where you focus your attention.

Questions for Reflection

1. What's an area of your life where your attention has shaped your reality—for better or worse?

2. How can you use the observer effect to achieve a specific goal or improve a relationship?

3. What beliefs or expectations do you hold that might be influencing your outcomes? How can you shift them to create a more positive reality?

This chapter uses the concept of the observer effect to explore the power of attention in shaping our experiences and outcomes. By drawing parallels between the quantum world and everyday life, it helps readers understand how their focus and expectations can influence reality, offering practical insights for harnessing this power to create positive change.

Chapter 7: Quantum Coherence—The Power of Alignment

The Symphony of Life

Imagine an orchestra tuning before a performance. Each musician plays their instrument independently, creating a cacophony of sounds. But when the conductor raises their baton, the musicians align, and the disparate notes merge into a harmonious symphony. This moment of alignment, where individual efforts come together to create something greater, is a beautiful metaphor forquantum coherence—a phenomenon where particles synchronize and work in unison. In our lives, coherence is the key to achieving harmony, whether in our personal goals, relationships, or teams.

What Is Quantum Coherence?

Quantum coherence occurs when particles, such as electrons or photons, exist in a synchronized state, behaving as a unified system rather than individual entities. This coherence allows for phenomena like superconductivity (where electricity flows without resistance) and quantum computing (where qubits work together to perform complex calculations). Coherence is fragile—it can be disrupted by external disturbances—but when maintained, it creates extraordinary efficiency and power.

Quantum Coherence in Everyday Life

Just as particles can synchronize to create coherence, humans can align their actions, thoughts, and energies to achieve greater harmony and effectiveness. Whether it's a team working toward a common goal, a family navigating challenges together, or an individual aligning their values with their actions, coherence is the foundation of success and fulfillment. This chapter explores how the principles of quantum coherence can help us create alignment in our lives.

The Science of Coherence

1.Superposition and Entanglement

Coherence relies on particles existing in superposition (multiple states at once) and entanglement (deep connections between particles). Similarly, human coherence requires openness to possibilities and strong connections with others.

2.Resonance and Synchronization

In coherent systems, particles resonate with each other, creating a synchronized state. In human systems, resonance can be seen in the way emotions, ideas, and actions align to create collective momentum.

3.Fragility and Maintenance

Quantum coherence is easily disrupted by external noise or interference. Similarly, alignment in human systems requires ongoing effort to maintain focus and harmony.

Real-Life Examples of Coherence

-High-Performing Teams: A sports team or a business team that works in perfect sync, anticipating each other's moves and aligning their efforts toward a shared goal.

-Healthy Relationships: A couple or family that communicates openly, supports each other's goals, and navigates challenges together.

-Personal Alignment: An individual whose actions, values, and goals are in harmony, leading to a sense of purpose and fulfillment.

The Quantum Mechanics of Alignment

Quantum coherence reminds us that alignment isn't just about individual effort—it's about creating a unified system where the whole is greater than the sum of its parts. Just as coherent particles work together to achieve extraordinary outcomes, aligned individuals and teams can achieve remarkable results.

Practical Takeaways: How to Create Coherence

1.Clarify Your Goals and Values

Just as particles need a shared state to achieve coherence, individuals and teams need a clear sense of purpose. Define your goals and values, and ensure they align with your actions.

2.Foster Strong Connections

Coherence relies on entanglement—deep, meaningful connections. Invest in relationships, communication, and trust to create a foundation for alignment.

3.Create Resonance

Align your actions, thoughts, and emotions to create a sense of flow and momentum. This might involve practices like mindfulness, visualization, or goal-setting.

4.Protect Your Coherence

Just as quantum coherence can be disrupted by external noise, human alignment can be disrupted by distractions or conflicts. Set boundaries, manage stress, and prioritize what matters most.

A Deeper Dive: Coherence in Nature

Quantum coherence isn't just a laboratory phenomenon—it's found in nature, too. For example, photosynthesis relies on coherent energy transfer in plants, allowing them to convert sunlight into energy with remarkable efficiency. This natural coherence is a reminder that alignment is a fundamental principle of life, not just physics.

Closing Thought: The Beauty of Coherence

Quantum coherence is a reminder that we are at our best when we are aligned—with ourselves, with others, and with the world around us. Just as particles synchronize to create extraordinary phenomena, we can achieve extraordinary outcomes when we work in harmony. So the next time you feel out of sync, remember: coherence isn't a distant ideal. It's a state you can create, one step at a time.

Questions for Reflection

1. What's an area of your life where you feel aligned—or out of alignment? What's contributing to that state?

2. How can you create greater coherence in your personal goals, relationships, or team dynamics?

3. What distractions or conflicts are disrupting your alignment, and how can you address them?

This chapter uses the concept of quantum coherence to explore the power of alignment in achieving harmony and success. By drawing parallels between the quantum world and human systems, it helps readers understand how to create coherence in their lives, offering practical insights for fostering alignment and achieving extraordinary outcomes.

Chapter 8: Entropy and Order—The Chaos of Everyday Life

The Never-Ending Laundry Pile

You've just finished folding a mountain of laundry, and for a brief moment, your closet looks pristine. But within days, the chaos returns—clothes are strewn about, socks go missing, and the cycle begins again. This constant battle between order and disorder isn't just a quirk of household chores; it's a reflection of a fundamental principle in physics:entropy. In the universe, entropy is the measure of disorder, and it always increases over time. But what does this mean for our lives, and how can we navigate the chaos?

What Is Entropy?

Entropy is a concept from thermodynamics that describes the tendency of systems to move from order to disorder. It's why ice melts, why buildings crumble, and why your desk inevitably becomes cluttered. In the quantum world, entropy reflects the number of ways particles can be arranged—the more possibilities, the higher the entropy. While entropy is often associated with chaos, it's also a source of creativity and change.

Entropy in Everyday Life

Just as physical systems tend toward disorder, our lives are constantly pulled between order and chaos. From the clutter in our homes to the unpredictability of our schedules, entropy is a force we grapple with daily.

But entropy isn't just a challenge—it's also an opportunity. This chapter explores how the principles of entropy can help us understand and embrace the chaos of everyday life.

The Science of Entropy

1.The Second Law of Thermodynamics

This law states that in any closed system, entropy always increases. While we can create local pockets of order (like cleaning a room), the overall trend is toward disorder.

2.Statistical Mechanics

Entropy is related to the number of possible states a system can be in. The more disordered a system, the more ways its particles can be arranged.

3.Entropy and Information

In information theory, entropy measures uncertainty or randomness. The higher the entropy, the more information is needed to describe the system.

Real-Life Examples of Entropy

-Household Chaos: No matter how often you clean, your home will eventually become messy again. This is entropy in action.
-Workplace Dynamics: Even the most organized teams face unexpected challenges and disruptions. Entropy reminds us that flexibility is key.

-Personal Growth: Growth often involves breaking down old patterns (increasing entropy) to create new, more adaptive ones (reducing entropy).

The Quantum Mechanics of Chaos

In quantum systems, entropy reflects the uncertainty and randomness of particle behavior. Similarly, in our lives, entropy represents the unpredictability and complexity we face. While this can feel overwhelming, it's also a source of potential. Just as entropy drives change in the universe, it can drive growth and transformation in our lives.

Practical Takeaways: How to Navigate Entropy

1.Embrace Imperfection

Entropy reminds us that perfection is impossible. Instead of striving for flawless order, focus on creating functional, adaptable systems.

2.Build Resilience

Chaos is inevitable, but resilience can help you navigate it. Develop strategies for managing stress, adapting to change, and recovering from setbacks.

3.Create Small Pockets of Order

While you can't eliminate entropy, you can create local areas of order. Focus on what you can control, whether it's a tidy workspace or a consistent routine.

4.See Chaos as Opportunity

Entropy isn't just a force of disorder—it's also a source of creativity and innovation. Embrace the unexpected, and look for opportunities in the chaos.

A Deeper Dive: Entropy and the Arrow of Time

Entropy is closely linked to the concept of time. In physics, the increase of entropy defines the "arrow of time"—the direction in which time flows. This means that entropy isn't just a measure of disorder; it's a fundamental feature of how we experience time and change.

Closing Thought: The Beauty of Entropy

Entropy is often seen as a negative force, a reminder of the inevitable decline into chaos. But it's also a source of creativity, growth, and transformation. Just as the universe uses entropy to drive change, we can use it to fuel our own evolution. So the next time you're faced with chaos, remember: entropy isn't your enemy. It's your ally in the journey of life.

Questions for Reflection

1. What's an area of your life where you feel overwhelmed by chaos? How can you create small pockets of order?

2. How do you typically respond to unexpected changes or disruptions? How can you build resilience to navigate entropy more effectively?

3. Can you think of a time when chaos led to growth or opportunity? What did you learn from that experience?

This chapter uses the concept of entropy to explore the balance between order and chaos in our lives. By drawing parallels between the quantum world and everyday experiences, it helps readers understand and embrace the inevitability of disorder, offering practical insights for navigating chaos and finding opportunity in the unexpected.

Chapter 9: Quantum Leaps—Moments of Sudden Change

The Lightning Bolt Moment

Imagine you're stuck in a rut, going through the motions of life, when suddenly—something shifts. Maybe it's a conversation that sparks a new idea, a setback that forces you to rethink your path, or a moment of clarity that changes everything. In an instant, your trajectory changes. This isn't just a stroke of luck; it's aquantum leap—a sudden, transformative shift that propels you into a new state of being.

What Are Quantum Leaps?

In quantum physics, a quantum leap refers to the abrupt transition of an electron from one energy level to another. Unlike the gradual changes we're used to in the macroscopic world, quantum leaps are instantaneous and discontinuous. They represent moments of sudden, profound change—and they have a powerful parallel in our personal lives. This chapter explores how the concept of quantum leaps can help us understand and embrace moments of transformation.

Quantum Leaps in Everyday Life

Just as electrons make sudden jumps between energy levels, humans experience moments of rapid change and growth. These quantum leaps can be triggered by external events—like a job loss, a new relationship, or a global crisis—or by internal shifts, such as a change in perspective or a breakthrough in self-awareness. This chapter delves into the science and psychology of quantum leaps, offering insights into how we can navigate and harness these moments of transformation.

The Science of Quantum Leaps

1.Energy Levels and Transitions

In atoms, electrons occupy specific energy levels. When an electron absorbs or releases energy, it can jump to a higher or lower level instantaneously. This discontinuous change is a hallmark of quantum systems.

2.Probability and Uncertainty

Quantum leaps are probabilistic—they can't be predicted with certainty, only described in terms of likelihood. This mirrors the unpredictability of transformative moments in our lives.

3.Nonlinear Growth

Quantum leaps challenge the idea that growth is always gradual and linear. Sometimes, progress happens in sudden bursts, reshaping our trajectory in unexpected ways.

Real-Life Examples of Quantum Leaps

-Career Shifts: A sudden opportunity or realization that leads to a new career path, even if it seems risky or unconventional.

-Personal Growth: A moment of insight or self-awareness that fundamentally changes how you see yourself and the world.

-Creative Breakthroughs: An artist or writer experiencing a sudden surge of inspiration, leading to a masterpiece.

-Relationships: A single conversation or event that transforms the dynamics of a relationship, for better or worse.

The Quantum Mechanics of Transformation

Quantum leaps remind us that change isn't always gradual or predictable. Just as electrons can jump between energy levels, we can experience sudden shifts that redefine our lives. These moments of transformation are often accompanied by uncertainty and discomfort, but they also offer the potential for profound growth and renewal.

Practical Takeaways: How to Navigate Quantum Leaps

1.Embrace Uncertainty

Quantum leaps are inherently unpredictable. Instead of fearing the unknown, embrace it as a source of possibility and growth.

2.Stay Open to Change

Transformation often requires letting go of old patterns and beliefs. Stay open to new ideas, experiences, and perspectives.

3.Recognize the Signs

Quantum leaps can be triggered by external events or internal shifts. Pay attention to moments of discomfort, insight, or inspiration—they may be the seeds of transformation.

4.Take Bold Action

When a quantum leap presents itself, don't hesitate. Take bold, decisive action to seize the opportunity and create lasting change.

A Deeper Dive: The Role of Energy in Quantum Leaps

In quantum physics, energy is the driving force behind leaps. Similarly, in our lives, energy—whether it's emotional, mental, or physical—plays a key role in transformation. By cultivating and channeling our energy, we can create the conditions for quantum leaps to occur.

Closing Thought: The Beauty of Quantum Leaps

Quantum leaps remind us that life is full of surprises. Just when we think we're stuck, the universe can offer a sudden shift that propels us forward. These moments of transformation are not just disruptions—they're opportunities to redefine ourselves and our paths. So the next time you feel the ground shifting beneath your feet, remember: you're not falling. You're leaping.

Questions for Reflection

1. What's a moment in your life that felt like a quantum leap? How did it change your trajectory?

2. How do you typically respond to sudden changes or disruptions? How can you embrace them as opportunities for growth?

3. What's an area of your life where you're ready for a quantum leap? What steps can you take to create the conditions for transformation?

This chapter uses the concept of quantum leaps to explore moments of sudden change and transformation in our lives. By drawing parallels between the quantum world and personal growth, it helps readers understand and embrace these moments, offering practical insights for navigating and harnessing the power of transformation.

Chapter 10: The Multiverse of You—Parallel Lives and Possibilities

The Road Not Taken

Imagine standing at a crossroads, faced with a life-altering decision. Do you take the job offer in a new city, or stay in your current role? Do you end a relationship, or commit to working through the challenges? Each choice leads to a different version of your life—a parallel universe where a different "you" walks a different path. This idea isn't just the stuff of science fiction; it's a concept rooted in quantum physics:the multiverse. This chapter explores how the multiverse can help us understand the infinite possibilities of our lives.

What Is the Multiverse?

The multiverse theory suggests that our universe is just one of countless parallel universes, each representing a different outcome of every possible event. In quantum physics, this idea arises from the many-worlds interpretation, which proposes that every quantum decision creates a branching of realities. While the multiverse remains a theoretical concept, it offers a powerful metaphor for the choices and possibilities we face in our lives.

The Multiverse in Everyday Life

Just as the multiverse represents infinite possibilities, our lives are shaped by the choices we make—and the paths we don't take. Every decision, no matter how small, creates a new version of reality. This chapter explores how the concept of the multiverse can help us reflect on our choices, embrace uncertainty, and imagine the infinite potential of our lives.

The Science of the Multiverse

1.Quantum Superposition and Branching

In quantum mechanics, particles exist in a superposition of states until observed. The many-worlds interpretation suggests that every possible outcome of a quantum event actually occurs, creating branching universes.

2.Parallel Realities

In the multiverse, every decision point leads to a new universe. This means that somewhere, a version of you made every possible choice.

3.Infinite Possibilities

The multiverse isn't just about alternate versions of your life—it's about the infinite potential of existence. Every possible outcome, no matter how unlikely, exists in some universe.

Real-Life Examples of the Multiverse

-Career Paths: The version of you who took that risky job offer might be living a completely different life than the version who played it safe.
-Relationships: The version of you who stayed in a relationship might be experiencing a different kind of growth than the version who walked away.

-Personal Growth: The version of you who pursued a passion might be living a life of creativity and fulfillment, while the version who ignored it might be stuck in a rut.

The Quantum Mechanics of Possibility

The multiverse reminds us that life is full of infinite possibilities. Just as every quantum decision creates a branching of realities, every choice we make opens up new paths and closes others. This can feel overwhelming, but it's also liberating. It means that no matter where we are in life, there are always new possibilities to explore.

Practical Takeaways: How to Navigate the Multiverse of You

1.Embrace Uncertainty

The multiverse is a reminder that life is inherently uncertain. Instead of fearing the unknown, embrace it as a source of possibility and growth.

2.Reflect on Your Choices

Take time to reflect on the paths you've taken and the ones you haven't. What can you learn from the choices you've made? What possibilities still lie ahead?

3.Imagine Alternate Realities

Use the concept of the multiverse to explore alternate versions of your life. What would the version of you in a parallel universe do? How can you bring some of that possibility into your current reality?

4.Focus on the Present

While it's fascinating to imagine parallel lives, the most important version of you is the one living right now. Focus on making choices that align with your values and goals.

A Deeper Dive: The Many-Worlds Interpretation

The many-worlds interpretation is one of the most intriguing ideas in quantum physics. It suggests that every quantum event creates a branching of realities, leading to an infinite number of parallel universes. While this remains a theoretical concept, it offers a profound perspective on the nature of reality and the power of choice.

Closing Thought: The Beauty of the Multiverse

The multiverse is a reminder that life is full of infinite possibilities. Every choice you make creates a new version of reality, and every path not taken exists in some parallel universe. This isn't just a scientific concept—it's a metaphor for the richness and complexity of existence. So the next time you're faced with a decision, remember: you're not just choosing a path. You're creating a universe.

Questions for Reflection

1. What's a decision in your life that led to a significant change? How might your life be different if you had chosen differently?

2. What's a path you didn't take that you still wonder about? How can you explore that possibility in your current life?

3. How can you use the concept of the multiverse to embrace uncertainty and explore new possibilities?

This chapter uses the concept of the multiverse to explore the infinite possibilities of our lives. By drawing parallels between the quantum world and personal choices, it helps readers reflect on their decisions, embrace uncertainty, and imagine the infinite potential of their existence.

Conclusion: Living in a Quantum World

Recap: The Quantum in the Quotidian

As we've journeyed through the pages of this book, we've explored how the strange and fascinating principles of quantum physics—superposition, entanglement, uncertainty, wave-particle duality, quantum tunneling, the observer effect, coherence, entropy, quantum leaps, and the multiverse—are not just abstract theories confined to laboratories. They are woven into the fabric of our everyday lives, shaping our decisions, relationships, challenges, and growth. From the moment you wake up to the choices you make, the connections you nurture, and the obstacles you overcome, the quantum world is there, offering insights and inspiration.

The Quantum Lens: A New Way of Seeing

By viewing life through a quantum lens, we gain a fresh perspective on the world and ourselves. We learn to:

-Embrace uncertainty as a source of possibility rather than fear.

-Celebrate our multifaceted nature, recognizing that we are not one thing but many.

-Harness the power of attention to shape our reality and create positive change.

-Navigate chaos and disorder with resilience and creativity.

-Seize moments of transformation, knowing that growth often happens in sudden leaps.

-Imagine infinite possibilities, understanding that every choice opens new paths.

This quantum perspective doesn't just deepen our understanding of the universe—it enriches our understanding of ourselves. It reminds us that life is dynamic, interconnected, and full of potential.

The Beauty of the Quantum Mundane

At the heart of this book is a simple yet profound idea: the mundane is anything but ordinary. The routines, decisions, and relationships that make up our daily lives are infused with the same principles that govern the behavior of particles and the fabric of reality. By recognizing the quantum in the mundane, we can find wonder, meaning, and inspiration in the everyday.

-Your morning routine is a dance of superposition and decision-making.
-Your relationships are a web of entanglement, defying distance and logic.
-Your struggles and triumphs are shaped by uncertainty, coherence, and quantum leaps.
-Your choices create a multiverse of possibilities, each one a new version of reality.

A Call to Curiosity

The quantum world invites us to be curious, to question, and to explore. It challenges us to see beyond the surface, to look for the hidden connections and patterns that shape our lives. Whether you're facing a difficult decision, navigating a relationship, or pursuing a dream, the principles of quantum physics offer a framework for understanding and growth.

-Ask questions: What possibilities exist in this moment? How can I embrace uncertainty as a source of creativity?
-Seek connections: How are my actions and choices influencing the world around me? How can I nurture the bonds that matter most?
-Embrace change: How can I use moments of chaos or transformation to create something new and meaningful?

Living in a Quantum World

Living in a quantum world means embracing the complexity, uncertainty, and interconnectedness of life. It means recognizing that we are not passive observers but active participants in shaping our reality. It means seeing the beauty in the mundane and the potential in the unknown.

As you close this book and return to your daily life, I invite you to carry this quantum perspective with you. Notice the moments of superposition in your decisions, the entanglement in your relationships, the coherence in your goals, and the leaps in your growth. See the world not as a fixed, predictable place but as a dynamic, ever-changing tapestry of possibilities.

Final Thought: You Are Quantum

You are not just a bystander in the universe—you are a part of it. Your thoughts, actions, and choices ripple through the fabric of reality, shaping the world around you. Just as particles exist in a state of potential, so do you. Every moment is an opportunity to create, to connect, and to grow.

So the next time you're faced with a challenge, a decision, or a moment of doubt, remember: you are quantum. You are superposition, entanglement, uncertainty, and possibility. You are the observer and the observed, the wave and the particle, the chaos and the order. You are a universe of potential, waiting to unfold.

Questions for Reflection

1. What's one quantum concept from this book that resonated with you the most? How can you apply it to your life?

2. How has this book changed the way you see the world and yourself?

3. What's a moment in your life where you can now see the quantum at work? How does this new perspective change how you feel about it?

This conclusion ties together the themes of the book, inviting readers to see their lives through a quantum lens and embrace the beauty, complexity, and potential of the everyday. It leaves them with a sense of wonder and empowerment, ready to explore the quantum world within and around them.

Glossary: Quantum Terms Made Simple

This glossary is designed to demystify the key quantum physics terms used throughout the book. Each term is explained in simple, relatable language, with examples from everyday life to make the concepts accessible and engaging. Whether you're new to quantum physics or just need a refresher, this glossary will help you navigate the quantum world with confidence.

A. Superposition

Definition: The ability of a quantum system (like an electron) to exist in multiple states or locations at once until it is observed or measured.

Everyday Example: Imagine you're deciding what to have for breakfast. Until you choose, you're in a superposition of all possible options—cereal, eggs, toast, or skipping breakfast altogether. The moment you decide, the superposition collapses into one reality.

Why It Matters: Superposition teaches us that life is full of possibilities, and our choices shape our reality.

B. Entanglement

Definition: A phenomenon where two or more particles become deeply connected, so that the state of one instantly influences the state of the other, no matter how far apart they are.

Everyday Example: Think of a close friendship where you can sense your friend's mood even when they're miles away. You might text them, only to find out they were thinking of you at the same time.

Why It Matters: Entanglement reminds us that our connections with others are powerful and enduring, transcending physical distance.

C. Uncertainty Principle

Definition: A principle stating that it's impossible to know both the position and momentum of a particle with absolute precision. The more you know about one, the less you know about the other.

Everyday Example: Imagine trying to plan your day down to the minute. The more precise you are with your schedule, the less flexibility you have to adapt to unexpected changes.

Why It Matters: The Uncertainty Principle teaches us to embrace uncertainty and focus on what we can control.

D. Wave-Particle Duality

Definition: The concept that particles, such as electrons and photons, can behave both as particles (discrete, localized entities) and as waves (spread-out, oscillating patterns), depending on how they're observed.

Everyday Example: Think of yourself at work versus at home. At work, you might be focused and task-oriented (particle-like), while at home, you're relaxed and emotionally attuned (wave-like). Both are authentic aspects of who you are.

Why It Matters: Wave-particle duality reminds us that we are multifaceted, with different roles and traits that emerge in different contexts.

E. Quantum Tunneling

Definition: A phenomenon where a particle passes through a barrier that, according to classical physics, it shouldn't be able to penetrate.

Everyday Example: Imagine trying to climb over a high wall to reach a goal. Quantum tunneling is like finding a hidden door that lets you pass through the wall instead.

Why It Matters: Quantum tunneling inspires us to find creative solutions and persevere, even when obstacles seem insurmountable.

F. Observer Effect

Definition: The idea that the act of observing or measuring a quantum system changes its state.

Everyday Example: Think of how your mood can shift when someone pays attention to you. Their observation influences your behavior, just as observing a particle changes its state.

Why It Matters: The Observer Effect reminds us that our attention and focus shape our reality, both for ourselves and others.

G. Quantum Coherence

Definition: A state where particles are synchronized and work together as a unified system, creating efficiency and harmony.

Everyday Example: Picture a team working seamlessly together, each member contributing their strengths to achieve a common goal.

Why It Matters: Quantum coherence teaches us the power of alignment—whether in teams, relationships, or personal goals.

H. Entropy

Definition: A measure of disorder or randomness in a system. In the universe, entropy always increases over time.

Everyday Example: Think of your desk. No matter how often you tidy it, it eventually becomes cluttered again. That's entropy in action.

Why It Matters: Entropy reminds us that chaos is a natural part of life, but we can create pockets of order to thrive.

I. Quantum Leap

Definition: A sudden, discontinuous change in the state of a quantum system, such as an electron jumping from one energy level to another.

Everyday Example: Imagine finally mustering the courage to quit a job you hate and pursue your passion. That moment of decision is a quantum leap.

Why It Matters: Quantum leaps inspire us to embrace sudden shifts and transformative moments in our lives.

J. Multiverse

Definition: The idea that our universe is just one of countless parallel universes, each representing a different outcome of every possible event.

Everyday Example: Think of a major life decision, like moving to a new city. In one universe, you took the leap and thrived. In another, you stayed and built a different kind of life.

Why It Matters: The multiverse reminds us that every choice opens new possibilities, and there's no single "right" path.

K. Wave Function

Definition: A mathematical description of the quantum state of a system, representing the probabilities of all possible outcomes.

Everyday Example: Imagine planning a vacation. The wave function is like a list of all possible destinations, activities, and outcomes. The moment you book your trip, the wave function collapses into one reality.

Why It Matters: The wave function illustrates the probabilistic nature of life and the power of choice.

L. Quantum Superposition

Definition: A specific application of superposition where a quantum system exists in a combination of all possible states until observed.

Everyday Example: Think of a coin spinning in the air. Until it lands, it's in a superposition of both heads and tails.

Why It Matters: Quantum superposition highlights the fluidity of reality and the potential for multiple outcomes.

M. Quantum Entanglement

Definition: A specific type of entanglement where particles remain connected even when separated by vast distances.

Everyday Example: Imagine two best friends who can finish each other's sentences, even when they're on opposite sides of the world.

Why It Matters: Quantum entanglement underscores the deep, invisible connections that bind us to others.

N. Quantum Fluctuations

Definition: Temporary changes in energy that occur at the quantum level, leading to the creation of particle-antiparticle pairs.

Everyday Example: Think of the random bursts of creativity or inspiration that seem to come out of nowhere.

Why It Matters: Quantum fluctuations remind us that even in emptiness, there's potential for creation and change.

O. Quantum Field

Definition: A field that permeates all of space, through which particles interact and exchange energy.

Everyday Example: Think of the air around you. You can't see it, but it's always there, allowing you to breathe and hear sounds.

Why It Matters: Quantum fields remind us of the invisible forces that shape our reality.

This glossary provides a clear, relatable introduction to the key concepts of quantum physics, making them accessible to readers of all backgrounds. By connecting these ideas to everyday experiences, it helps readers see the quantum world not as an abstract theory, but as a lens for understanding their own lives.

Practical Exercises: Applying Quantum Principles to Your Life

These hands-on exercises are designed to help readers take the abstract concepts of quantum physics and apply them to their everyday lives. Each exercise is tied to a specific quantum principle, offering a tangible way to explore and internalize the ideas discussed in the book. These activities are simple, actionable, and transformative, encouraging readers to see their lives through a quantum lens.

A. Superposition Journaling

Quantum Principle: Superposition

Objective: Explore the power of possibilities and decision-making.

Exercise:

1. Think of a decision you're currently facing (e.g., a career move, a relationship choice, or a personal goal).

2. Write down all the possible outcomes as if they're all happening at once. For example, if you're deciding whether to take a new job, describe what your life would look like if you took the job, if you stayed in your current role, or if you pursued a completely different path.

3. Reflect on how it feels to hold space for multiple possibilities. Does it make the decision easier or harder? Does it open your mind to new options?

Takeaway: This exercise helps you embrace the idea of superposition, recognizing that life is full of possibilities and that your choices shape your reality.

B. Entanglement Mapping

Quantum Principle: Entanglement

Objective: Visualize and nurture your deep connections with others.

Exercise:

1. Draw a diagram (or use a mind-mapping tool) with yourself at the center.

2. Add branches for the people in your life who you feel deeply connected to—family, friends, mentors, or colleagues.

3. For each person, write a few words about how they influence your thoughts, emotions, or actions.

4. Reflect on how these connections shape your life. Are there any relationships you'd like to strengthen or let go of?

Takeaway: This exercise helps you appreciate the power of entanglement and the invisible threads that bind you to others.

C. Uncertainty Reflection

Quantum Principle: Uncertainty Principle

Objective: Embrace uncertainty as a source of growth and opportunity.

Exercise:

1. Make a list of things in your life that feel uncertain or unpredictable (e.g., your career path, a relationship, or your health).

2. For each item, write down how you typically respond to the uncertainty (e.g., with fear, excitement, or avoidance).

3. Now, reframe each uncertainty as an opportunity. For example, if you're unsure about your career, write about how this uncertainty could lead to new possibilities.

4. Choose one area of uncertainty and take one small step to explore it further (e.g., research a new career path or have an open conversation with a loved one).

Takeaway: This exercise helps you shift your mindset from fearing uncertainty to seeing it as a source of potential.

D. Wave-Particle Duality Role Play

Quantum Principle: Wave-Particle Duality

Objective: Explore the different roles and identities you embody.

Exercise:

1. Write down the different roles you play in life (e.g., parent, professional, friend, artist, student).

2. For each role, describe how you behave, think, and feel when you're in that mode.

3. Reflect on how these roles complement or conflict with each other. Are there any roles you'd like to integrate more fully into your life?

4. Choose one role and spend a day intentionally embodying it. Notice how it feels to focus on that aspect of yourself.

Takeaway: This exercise helps you celebrate your multifaceted nature and explore how different roles shape your identity.

E. Quantum Tunneling Visualization

Quantum Principle: Quantum Tunneling

Objective: Find creative ways to overcome obstacles.

Exercise:

1. Identify a barrier or challenge in your life (e.g., a fear, a limiting belief, or a practical obstacle).

2. Close your eyes and visualize the barrier as a solid wall. Imagine yourself trying to climb over it, push through it, or go around it.

3. Now, imagine a hidden door or tunnel in the wall. Visualize yourself passing through it and emerging on the other side, free from the obstacle.

4. Write down one creative solution or action step that could help you "tunnel through" the barrier in real life.

Takeaway: This exercise inspires you to think outside the box and find innovative ways to overcome challenges.

F. Observer Effect Experiment

Quantum Principle: Observer Effect

Objective: Explore how your attention shapes your reality.

Exercise:

1. Choose an area of your life where you'd like to see change (e.g., your mood, a relationship, or a habit).

2. Spend a week intentionally observing and focusing on that area. For example, if you want to improve your mood, pay close attention to what makes you happy or stressed.

3. At the end of the week, reflect on how your attention influenced your experience. Did focusing on the area lead to any changes or insights?

Takeaway: This exercise helps you harness the power of the observer effect, showing how your attention can shape your reality.

G. Coherence Meditation

Quantum Principle: Quantum Coherence

Objective: Create alignment and harmony in your life.

Exercise:

1. Find a quiet space and sit comfortably. Close your eyes and take a few deep breaths.

2. Visualize your life as a system of interconnected parts—your goals, relationships, values, and actions.

3. Imagine each part coming into alignment, like particles in a coherent state. Feel the sense of harmony and flow that comes from this alignment.

4. Set an intention to bring this coherence into your daily life. For example, you might decide to align your actions with your values or create more harmony in your relationships.

Takeaway: This exercise helps you cultivate coherence, creating a sense of alignment and purpose in your life.

H. Quantum Leap Goal Setting

Quantum Principle: Quantum Leaps

Objective: Prepare for sudden shifts and transformative moments.

Exercise:

1. Identify an area of your life where you're ready for a sudden shift (e.g., your career, health, or creativity).

2. Write down what a "quantum leap" in that area would look like. For example, if you're focusing on your career, your leap might be landing your dream job or starting your own business.

3. List one bold action you can take to create the conditions for that leap (e.g., updating your resume, networking, or taking a class).

4. Commit to taking that action within the next week.

Takeaway: This exercise helps you prepare for and embrace moments of sudden transformation.

I. Multiverse Imagination

Quantum Principle: Multiverse

Objective: Explore the infinite possibilities of your life.

Exercise:

1. Think of a major decision you've made in the past (e.g., moving to a new city, ending a relationship, or choosing a career path).

2. Imagine the alternate versions of your life that might exist in parallel universes. Write a short description of each version.

3. Reflect on how these alternate realities make you feel. Are there any aspects of those versions that you'd like to bring into your current life?

Takeaway: This exercise helps you see the richness of your life's possibilities and appreciate the path you've chosen.

These practical exercises provide readers with tools to apply quantum principles to their own lives, fostering personal growth, self-awareness, and creativity. By engaging with these activities, readers can deepen their understanding of the quantum world and see its relevance in their everyday experiences.

Interactive Quantum Experiments

These interactive experiments are designed to bring the principles of quantum physics to life in a fun, hands-on way. While some of these experiments are thought experiments (exercises you can do in your mind), others involve simple materials you can find at home. Each experiment is tied to a specific quantum concept, helping readers experience the strange and fascinating world of quantum physics firsthand.

A. Double-Slit Thought Experiment

Quantum Principle: Wave-Particle Duality

Objective: Explore how particles can behave as both waves and particles.

Materials Needed: None (this is a thought experiment).

Instructions:

1. Imagine a barrier with two slits in it, like a piece of paper with two vertical cuts.

2. Picture shining a light through the slits. If light were purely a particle, you'd expect to see two bright spots on the wall behind the barrier.

3. Instead, imagine seeing an interference pattern—a series of light and dark bands—on the wall. This pattern is created because light behaves like a wave, with peaks and troughs interfering with each other.

4. Now, imagine placing a detector near the slits to observe which slit the light passes through. The interference pattern disappears, and you see two bright spots instead. This is because the act of observation forces the light to behave like a particle.

Takeaway: This experiment illustrates the strange nature of wave-particle duality and the role of observation in shaping reality.

B. Coin Entanglement Simulation

Quantum Principle: Entanglement

Objective: Simulate how entangled particles behave.

Materials Needed: Two coins (or any two identical objects).

Instructions:

1. Flip both coins simultaneously.

2. Observe the results. In a quantum entanglement scenario, the coins would always land on the same side (both heads or both tails), no matter how far apart they are.

3. Repeat the experiment several times, imagining that the coins are "entangled" and always mirror each other.

Takeaway: This simple simulation helps you understand the concept of entanglement and the idea that particles can be deeply connected, even at a distance.

C. Quantum Tunneling Visualization

Quantum Principle: Quantum Tunneling

Objective: Visualize how particles can pass through barriers.

Materials Needed: A piece of paper and a pen.

Instructions:

1. Draw a barrier (a thick line) on the paper. On one side of the barrier, draw a dot representing a particle. On the other side, draw a target.

2. Imagine the particle trying to reach the target. In classical physics, the particle would need enough energy to go over the barrier.

3. Now, imagine the particle "tunneling" through the barrier, appearing on the other side without going over it.

4. Reflect on how this concept applies to your life. Are there any barriers you're facing that might require a "quantum tunneling" approach?

Takeaway: This visualization helps you understand the concept of quantum tunneling and inspires creative problem-solving.

D. Observer Effect with a Flashlight

Quantum Principle: Observer Effect

Objective: Explore how observation can change a system.

Materials Needed: A flashlight and a dark room.

Instructions:

1. Turn off the lights and shine the flashlight on a wall. Notice how the light behaves when you're not directly observing it (e.g., the beam's shape and intensity).

2. Now, place your hand in front of the flashlight and observe how the light changes. The act of observing (by placing your hand in the beam) alters the system.

3. Reflect on how this experiment mirrors the observer effect in quantum physics, where the act of measurement changes the state of a particle.

Takeaway: This experiment helps you understand how observation can influence reality, both in quantum systems and in your own life.

E. Quantum Superposition with a Spinning Coin

Quantum Principle: Superposition

Objective: Explore the idea of multiple states existing at once.

Materials Needed: A coin.

Instructions:

1. Spin the coin on a flat surface. While it's spinning, it's in a superposition of both heads and tails.

2. Observe the coin as it slows down and eventually lands on one side. The superposition collapses into a single state (heads or tails).

3. Reflect on how this mirrors the concept of superposition in quantum physics, where particles exist in multiple states until observed.

Takeaway: This experiment helps you understand the probabilistic nature of quantum systems and the role of observation in collapsing possibilities into reality.

F. Quantum Fluctuations with a Bubble Wrap

Quantum Principle: Quantum Fluctuations

Objective: Visualize the random, unpredictable nature of quantum fluctuations.

Materials Needed: A sheet of bubble wrap.

Instructions:

1. Hold the bubble wrap in your hands and observe the bubbles. Each bubble represents a quantum fluctuation—a temporary change in energy.

2. Pop a few bubbles at random. Notice how the act of popping one bubble doesn't affect the others, but the overall system changes.

3. Reflect on how this mirrors the concept of quantum fluctuations, where particles and energy appear and disappear unpredictably.

Takeaway: This experiment helps you visualize the randomness and unpredictability of quantum fluctuations, which are a fundamental part of the quantum world.

G. Multiverse Imagination Exercise

Quantum Principle: Multiverse

Objective: Explore the idea of parallel universes and infinite possibilities.

Materials Needed: None (this is a thought experiment).

Instructions:

1. Think of a major decision you've made in your life (e.g., moving to a new city, choosing a career, or ending a relationship).

2. Imagine the alternate versions of your life that might exist in parallel universes. For example, in one universe, you took the job offer in another city. In another, you stayed in your hometown.

3. Write a short description of each alternate reality. How do they differ from your current life? What might you learn from these alternate versions of yourself?

Takeaway: This exercise helps you explore the concept of the multiverse and appreciate the infinite possibilities of your life.

H. Quantum Coherence with a Tuning Fork

Quantum Principle: Quantum Coherence

Objective: Explore the idea of synchronization and alignment.

Materials Needed: Two tuning forks (or a tuning fork and a resonant surface, like a table).

Instructions:

1. Strike one tuning fork and hold it near the second one. Notice how the second tuning fork begins to vibrate in sync with the first.

2. Reflect on how this mirrors the concept of quantum coherence, where particles synchronize and work together as a unified system.

3. Think about areas of your life where you'd like to create more coherence (e.g., aligning your goals, relationships, or values).

Takeaway: This experiment helps you understand the power of alignment and synchronization, both in quantum systems and in your own life.

These interactive experiments provide a hands-on way to explore the strange and fascinating world of quantum physics. By engaging with these activities, readers can deepen their understanding of quantum concepts and see their relevance in everyday life. Whether you're flipping coins, spinning a penny, or imagining parallel universes, these experiments make quantum physics accessible, engaging, and fun.